HUNCH

Kate Kennedy

HUNCH

OBERON BOOKS
LONDON

WWW.OBERONBOOKS.COM

First published in 2018 by Oberon Books Ltd
521 Caledonian Road, London N7 9RH
Tel: +44 (0) 20 7607 3637 / Fax: +44 (0) 20 7607 3629
e-mail: info@oberonbooks.com
www.oberonbooks.com

A catalogue record for this book is available from the British Library.

PB ISBN: 9781786826190
E ISBN: 9781786826138

Cover design by Pencil Bandit

eBook conversion by Lapiz Digital Services, India.

Hunch was first performed at Soho Theatre where it previewed 23 – 24 Jul 2018 before its run at Assembly Roxy, Edinburgh Fringe Festival 1 – 27 Aug 2018 with the following cast and creative team:

Writer & Performer	Kate Kennedy
Dramaturg & Director	Sara Joyce
Producer	Milly Smith at DugOut Theatre
Lighting	Jack Weir
Sound	Max Perryment
Design Concept	Max Dorey
Design	Natalie Johnson
Production Manager	Sara Hooppell
Photography	Richard Lakos
Grapic Design & Poster	Pencil Bandit
Stage Manager	Gemma Scott

Sara Joyce Director

Sara most recently directed *Dust* by Milly Thomas and *Elsa* by Isobel Rogers. Other directing credits include: *Split* by Emma Pritchard and Tamar Broadbent, *On Raftery's Hill* by Marina Carr at Bunker Theatre (reading) as part of Damsel Develops; *The Scar Test* by Hannah Khalil; Bush Theatre Emerging Writer's Showcase; Soho Young Playwright's Award Showcase; *Crestfall* by Mark O'Rowe; and *Act Without Words I, Rough For Theatre II, Catastrophe* by Samuel Beckett (Offie nominated for Best Director).

Sara has also directed *Click* and *The Win Bin* by Kate Kennedy, this will be their third collaboration.

Sara is a director with the Old Vic 12. She was resident director at Almeida Theatre and resident assistant director at Soho Theatre. She has worked as associate and assistant director with Dominic Dromgoole, Claire van Kampen, Gavin Quinn (Pan Pan Theatre co.) Richard Eyre, Rupert Goold and Steve Marmion.

She was shortlisted for the KSF Emerging Artist's Award and is a recipient of the Deutsche Bank Award for Creative Enterprises. She studied Drama and Theatre at Trinity College, Dublin and trained at Ecole Jacques Lecoq.

Milly Smith Producer

Milly is a theatre producer and manager, working in the West End and on tour within the UK and internationally.

She trained at The Oxford School of Drama before going to University in Leeds where she met George and joined DugOut as their producer.

She formerly worked for Jamie Hendry Productions, shows include: *Let It Be* (West End); *Neville's Island* (West End); and *Impossible* (West End/UK Tour/International).

For DugOut Theatre: *Songlines, Hunch, Replay, Swansong, Stack, Goodbear, The Sunset Five* and *Inheritance Blues*.

Film credits include: *All The Devils Men* (assistant production coordinator) and *Justice League* (assistant production coordinator).

Jack Weir Lighting Design

Jack trained at The Guildhall School and won the ETC award for Lighting Design in 2014.

In 2016 he was nominated for the WhatsOnStage award for Best Lighting Design for *The Boys In The Band* at the Vaudeville Theatre, West End. He was also double-nominated and a finalist in the 2016 Off West End Awards for Best Lighting Designer.

Recent work includes: *West Side Story* (Bishopsgate Institute); *Crazy For You* (Trinity Laban); *Grindr The Opera, Beautiful Thing* (Above The Stag); *Little Women* (Chichester Uni); *Pippin* (Tring Park School); *Rothschild & Sons* (Park Theatre); *Hanna* (Arcola Theatre/UK Tour); *Georges Marvellous Medicine* (Leicester Curve/Rose Theatre/UK Tour); *Pyar Actually* (Watford Palace/Tour); *Talk Radio* (Old Red Lion); *Holding The Man* (Above The Stag); *Salad Days* (Union Theatre); *Dust* (Soho Theatre); *Summer In London* (Theatre Royal Stratford); *Judy!* (The Arts Theatre); *Betty Blue Eyes* (Chichester Uni); *Assata Taught Me* (Gate Theatre); The *Plague* (Arcola Theatre); *Out Of Order* (Yvonne Arnaud/UK Tour); *Pray So Hard For You* (Finborough Theatre); *La Ronde* (Bunker Theatre); and *The Boys In The Band* (Vaudeville, West End).

Max Perryment Sound Design

Max is a London, UK based composer and sound designer. In addition to theatre he writes music for film, television commercials and dance. Max has an MA in Electroacoustic Composition from City University.

Recent theatre work includes: *Utility* (Orange Tree Theatre); *Rasheeda Speaking* (Trafalgar Studio 2); *Twilight: Los Angeles 1992* (Gate Theatre – Off West End Award Nomination); *Dust* (Soho Theatre/Trafalgar Studio 2); *The Tide* (Young Vic, YV Taking Part); *Romeo and Juliet* (Orange Tree Theatre); *Start Swimming* (Young Vic); and *Hair* (Hope Mill Theatre/The Vaults)

maxperryment.co.uk

Max Dorey Concept Deisgn

Max was nominated for Best Set Design for the 2016 UK Theatre Awards for *And Then Come The Nightjars*. He is also eight time nominated for Best Set Design in The Off West End Awards.

Credits include: *Teddy* (Vaults Theatre); *Julius Caesar: First Encounters* (Royal Shakespeare Company); *Talk Radio* (Old Red Lion); *Tryst* (Tabard Theatre); *The End of Hope* (Soho Theatre); *Insignificance* (Arcola Theatre); *Abigail* (Bunker Theatre); *Luv* (Park Theatre); *The Collector* (The Vaults); *Cargo* (Arcola Theatre); *Last of the Boys* (Southwark Playhouse); *After Independence* (Arcola Theatre/Papatango); *P'Yongyang* (Finborough Theatre); *No Villain* (Trafalgar Studios); *All The Little Lights* (Fifth Word); *And Then Come The Nightjars* (Theatre503/Bristol Old Vic/UK Tour); *Orson's Shadow, Teddy* (Southwark Playhouse); *Lardo, Marching On Together* (Old Red Lion); *Coolatully* (Finborough Theatre); *Sleight and Hand* (Edinburgh Fringe); *I Can Hear You, This Is Not An Exit* (The Place at the Courtyard, RSC/Royal Court Upstairs); *Count Ory* (Blackheath Halls); *Black Jesus* (Finborough Theatre); *The Duke In Darkness, Marguerite* (Tabard Theatre); and *Disco Pigs* (Tristan Bates).

Natalie Johnson Designer

Natalie trained at The Liverpool Institute for Performing Arts and was awarded the Liverpool Everyman and Playhouse Prize for Stage Design in 2017.

Recent credits as designer include: *To Anyone Who Listens* (Hen and Chickens, London); *The Wasp* (Hope Mill Theatre, Manchester); *Othello* (Everyman Theatre, Liverpool); and *Conversations With Rats* (theSpace On The Mile, Edinburgh).

As assistant designer credits include: *The Big I Am* (Everyman Theatre, Liverpool); *A Clockwork Orange* (Everyman Theatre, Liverpool); and *Paint Your Wagon* (Everyman Theatre, Liverpool).

Sara Hooppell Production Manager

Sara trained in classical theatre at the Rose Theatre, Kingston-upon-Thames. She is the co-founder of female-led collective VOLTA, regional touring company Fathom Theatre and is an associate artist of Coppice Theatre. Alongside her work as a production manager, Sara is an actor and producer, most recently working with Les Enfants Terribles, Four of Swords, FFP New Media and the Bike Shed Theatre.

Gemma Scott Stage Manager

Gemma has been a stage manager since 2011, working on everything from children's theatre to burlesque.

Recent credits include: *Truth* (Birmingham Rep/Song Theatre); *Raisins to Stay Alive* (Soho Theatre/zazU); *Educating Rita* (Queen's Theatre, Hornchurch); *Experience* (Hampstead Theatre); and *Tonight With Donny Stixx* (The Bunker/Metal Rabbit Productions).

DUGOUT
THEATRE

"This group bristles with talent." (The Stage)

Founded by a group of comedians, writers, musicians and actors, DugOut's work is often funny, sometimes sad, usually hopeful and almost always musical. It warms your heart, moves your feet, treats your eyes, lightens your load, tickles your fancy and swells your heart.

Taking our joyful, popular plays around the country, DugOut have produced a total of five hit shows to date – *Inheritance Blues*, *Fade*, *The Sunset Five*, *Swansong* and *Replay* and believe in bringing our unique brand of theatre magic to audiences everywhere.

"Full of vim and great music" (The Guardian)

DugOut shows have appeared at the Theatre 59E59 – New York, Sheffield Crucible, Soho Theatre, Arcola Theatre, HighTide Festival, West Yorkshire Playhouse, Hull Truck, Greenwich Theatre and The Pleasance (Edinburgh and London).

"Oozing with simple British charm" (Daily Mail)

You have brains in your head.
You have feet in your shoes.
You can steer yourself any direction you choose.

– Dr. Seuss

THANKS:

To Joycey, the 'gimlet eye' who believes anything is possible. May the free Pret coffees continue to flow and fuel your utter genius. Hum wouldn't be the same without you, thank you to the glittery moon and back.

To Mum, Dad, Alex, Nick, India and Jazz. The best superheroes I've ever met, all of you. An inch of your integrity and humour would annihilate a Marvel blockbuster.

To my tremendous mates who keep coming to see/read these things and don't whack their head on the table when they hear me say 'they said I'm down to the final two' with my fingers crossed and a wincey face. 'Being sound' is an understatement.

Milly Smith for her faith and brilliance, George Chilcott for more than a decade of friendship and the call up, Pencil Bandit and Richard Lakos for a belter of a poster, Max Perryment, Max Dorey, Jack Weir, Gemma Scott, Chloe Nelkin, Tilly Wilson, Camilla Young, Sarah MacCormick, Emma Higginbottom, Josh Byrne, Emma Harvey, Stewart Pringle, Peggy Ramsay, Captain Birdseye's creamed spinach, Breffney Cogan, Malcolm Gladwell, Dr. Seuss, Brainchild, Soho Theatre, Assembly Roxy and BDE.

*This text was sent to press during rehearsals
and may differ from what is performed on stage.*

1. HUNCHCHELLA PART ONE

HUNCH is waiting backstage.

Music.

HUNCH accepts her award on stage.

HUNCH: I have no words.
Words are…

HUNCH opens her mouth widely to try and summon words.

Me. Me. Me.
There's not many heroes who can say they've saved Hum and everyone in it.
Fasten your seatbelts, civilians. It's Hunch.

She does the HUNCH symbol.

2. HUNCHILDA

A year earlier. UNA's twenty-seventh birthday.

UNA: I hate occasions.
I know normal days. I know where everyone is, where they're about to go, what time I can go to sleep. Occasions are pranks.
Birthdays are bullies.

UNA, zero years old, being born in hospital.

MUM: Why she not coming out, Larry? Pull her out with a fucking crane if you have to!

DAD: Oh she's come out love. Oh, she's in, she's out. Oh. She's in.

MUM: This is a home birth Larry, not The Vagina Open.

DAD: Love, she's stopped halfway.

UNA: The first five years were full of too much learning. As soon as I'd get the hang of something, I was either 'too big' or 'too old' so it had to *(shudder)* change.

UNA, six years old. MUM whispers to NATHAN, UNA's brother.

MUM: *(To NATHAN.)* You might think that the claw will grab the toy, but it's an illusion Nathan, okay?

(To UNA.) Your brother is taking you birthday bowling, Una!

In bowling alley.

NATHAN: Just pick a ball, Una. They're all the same – just different colours. Grab, poke, chuck – *(Strike.)* Wicked.

UNA can't pick.

Just knock one over or Mum'll make me take you to the cinema.

UNA really can't pick. She panics over the choice. Sees the pins. Runs towards them. Fiercely kicks them one by one.

Woah – woah – woah – UNA!

UNA: School, I got used to. Wore the same thing everyday. Slept in my uniform. Ate only browns and greens.

Routine was my soulmate. And Jools.

JOOLS: Happy 13th Birthday Una Balloona! Make a wish!

UNA can't decide.

Oh. That's okay, I'll do it. Let's wish Mark from 7b discovers a razor for his bumfluff.

UNA nods.

Happy 16th Birthday Una Balloona! Make a w–!

UNA can't decide and her tummy hurts.

Woah – are you alright? I'll have it too and then we can both go home early.

(Phone message.) Happy 18th Birthday Una!
Sorry I'm late – I got a bit busy, with a male. I'll be there in 12 minutes. Love you Una.

UNA: With change came chronic pain. Any exotic change to my routine, any new discovery, surprise or choice made my tummy cramp.

UNA is twenty-seven years old talking to SAM, JOOLS' kid, from the treehouse.

Did you check?

SAM: Yes. Your mum, my mum, two stupid boys. A round yum and one present.

UNA: Nobody else there?

SAM: Nope.

UNA: Good work soldier.

SAM: Why do you have a treehouse when you're so old?

UNA: I only look old because I'm very tall, Sam. I'm actually what's considered quite young.

She's stuck.

Can you help me down?
Wait. Sam.
You need to take a step to the left.

Bird shit.

SAM: Woah! A bird plop! Are you magic?

They knock on the window.

Can we come in yet?

MUM/DAD/JOOLS: Sssssshhhhhhhhhhhhhhh/she's here, she's
here/right I'll hide the presents –

SAM: Do they know we can see them?

UNA: Oh yeah, it's the routine. *(Pretend.)* I wonder where they
all are?

The door opens.

NATHAN: HAPPY BIRTHDAY UNA! Twenty-seven candles.
Make a wish!

UNA looks pained. MUM punches NATHAN on the arm.

Ow!! Sorry, I mean, make a fish?

Watching TV.

UNA: The afternoon crew left (bye, thank you for the organised
fun!) I think I'm home dry.
Mum sidles up to me.

MUM: Your dad and I were wondering – would you go and
pick up…dinner tonight?

UNA: Pick up dinner?

MUM: Yes.

UNA: *(Sceptical.)* Okay. I can go and pick up dinner. What shall
I get?

MUM: You decide.

UNA: Like what?

MUM: You decide love.

UNA: No, no, you tell me what to get and I'll get it.

MUM: *(Losing her temper.)* Just pick. There is no wrong decision.
Your dad will eat anything.

DAD: Except oinks – they can hear us!

MUM: Larry.

UNA: Please just pick. I'll do it another day, not today, not my
birthday.

MUM: Una. We're only as good as the decisions we make. You
can do it.

3. HUGEST HUNCH YET

*In Hum's butchers. UNA attempts to get the dinner. She strains at the
multitude of choice behind the counter.*

UNA: Getting dinner.
Get-ting. Din-ner. Gettindinner.
Excuse me – do you have a special?
Never mind.

She looks at all the cases of food stuffs.

Excuse me – how many moos would I need for three
people? Actually – you can go in front.

*She watches the person in front order like a normal person and
then leave.*

Yeah, I'll have the same as them – three cluck wings. Two
quack legs, no four baa chops – no – um – do you want to
go ahead?

She lets another person go in front. Steps aside.

Sorry – just curious – what would you get for your own
birthday dinner –

BUTCHER: Are you in the queue this time, or…?

UNA: Um – I don't know. Yes. No. I don't know.

LADY WITH DOG: We know what we want, don't we Juju?

JUJU: Woof.

UNA: Um…

BUTCHER: If you need a recommendation, the organic oink is flying off the counter today –

She clutches her stomach. It's not a pain. It's a force.

Bugger me, if you're up the duff you can't be having it here.

UNA: No, I'm fine. It's a stitch.

BUTCHER: From queueing? The state of young people nowadays…

UNA: I'll have seven moos, eight moos – no – I'll just see if –

UNA looks as if she's having a child.

BUTCHER: Bloody hell.

The force is huge. It drags her towards the door of the butchers. She stops. Then –

UNA: Sorry. I'm sorry. Everyone should leave. Everyone needs to leave.

Civilians in the queue start to leave. UNA goes up to LADY WITH DOG.

Excuse me, you have to get out of here. You and your dog have to leave.

LADY WITH DOG: Don't look at the crazy lady JuJu. We'll have six oinks… Juju!

UNA eeks the sausage dog out of the lady's hands, carries it out the door. She's outside.

SMALL EXPLOSION.

She freezes.

BIGGER EXPLOSION.

Using Juju to cover her head she hits the ground. She looks towards the butchers.

UNA: Fuck me.

Cameras flash. Journalists burst out of nowhere. Paparazzi.

JOURNALIST 1: Excuse me can we have a few words? How did you know to leave the shop?

UNA: Um…

JOURNALISTS: What made you leave the shop?/Did you get a tip off?/Do you know the bomber?/Would you say you are involved in an extremist selective?/Are you the victim of blackmail?

UNA: Um. I felt it.

JOURNALIST 1: Where? Where did you feel it?

UNA: I've always felt it, a pain.

JOURNALIST 1: WHERE? WHERE?!

UNA: HERE! My tummy. Right here.

JOURNALIST 1 steps over a body.

JOURNALIST 1: And there we have it, a young woman who trusted her gut to survive…

4. THE TRUNK

Back at MUM and DAD's. UNA is in the bath.

MUM: Is it hot enough?

DAD: Pauline.

MUM: Well she won't say will she? Could be sub-zero and then we'd have a frozen lump of limbs.

UNA: It's fine. I'm fine.

She gets into the bath. It's freezing.

You can leave me alone.

Most civilians my age don't spend so much time with their parents, but mine aren't parents, they're my pals. Pals that had to do it for me to exist.

An automated message from The Trunk requesting her presence. She gets out of the bath.

Mum! Dad! It's The Trunk. They want to see me.
Dad's face drops. He's a stickler for authorit– authortiatar– hierarchy.

DAD: Do they think you were involved in the incident? Are they coming here?! Quick – Pauline, fetch the good tea towels.

UNA: They want to meet me this week.

MUM: Why would they want to meet you? They've got enough going on as it is.

DAD: Did they say anything about us? Do I have time to give my head a polish?

MUM: No offence love, but I bet its protocol. Maybe they're doing a civilian case study? Or random civilian checks?

UNA: They said they needed me.

MUM: Well, if they need you.

DAD: They must need you.

In front of the TV.

UNA: I pretended to fall asleep in front of the telly.

MUM and DAD whisper.

MUM: Larry. Do you think it might be a boy?

DAD: Unlikely, love.

MUM: I used to pull The Trunk card when I met you –

DAD: Did you now…

UNA: I'm awake. I'm awake!

5. THE TESTS

UNA is in a clinic. LUKE is sitting next to her.

UNA: They want to get me checked. For everything.

I've had my outsides scanned, my limbs measured, my skin scraped, my insides prodded, my ancestry analysed, my strength and stamina examined and the final test was – *sexual.*

NURSE: Swabby swabby.

UNA: This room is half-full of shifty eyes, praying they don't bump into their ex, and half-full of smug eyes, basking in post-bonk glory.

I'm not sure whether The Trunk can read my thoughts, so I think about the floor. Then the door. Then the chair. This chair. Or that chair. This chair.

I wonder if anyone has ever said no to The Trunk. Stop it Una, this chair. That chair.

LUKE arrives.

I recognise that bloke from Jools' house. He won't remember me.

LUKE: You taking that chair?

UNA: Oh. Yeah.

LUKE: So, you're too sexy for your own good too, eh? Take me down officer!

UNA: *(Deadpan.)* No I'm here to get my sexual health tested.

His cheeks flush red. I let out a delayed '*guffaw*' to show I didn't think he was weird. Didn't seem enough.

But *I'm* also too sexy *for* my own good.

I got all the intonation wrong in order for that to be humorous.

LUKE: We met at Jools' engagement liquids – I'm the work guy that had his tie tied around his head. Got the limbo going with um – her daughter um –

UNA: Sam!

LUKE: Yeah! I always meant to ask Jools to ask you for a naughty liquid out and about, is that something that would be up your boulevard?

UNA: Um – I – I – don't know.

My tummy moans.

'Luke to Room 47' flashes on the screen.

LUKE: Okay. Look. Monday. 7. The bar on the corner of North Hum.

UNA: Uh. Okay.

LUKE: And in the meantime – stop having SO MUCH SEX!

6. THE INITIATION

UNA is at the Superhero Headquarters outside the office of HEAD (JIM) and HEART (ROBBIE).

MAUDE: Need you to sign here, here and here.

UNA: I'm at The Trunk Headquarters and Maude, the secretary, is breathing very heavily out of her throat.

MAUDE: This is your oath to secrecy. You tell a soul and we take your soul. Any questions?

UNA: Um – do you have a bucket? Think I might blow.

MAUDE: *(She winces.)* Ew. They'll see you now.

UNA: The glare is so bright. I can just about make out three blobs.

Blob One grabs my arm.

HEAD: It'll take approximately twenty seconds for your eyes to adjust. We won't wait.

UNA: Sounds like Head – The Trunk's founder. Dad's a mega fan of Head – he's science and logic.

HEAD: So you'll be pleased to know your tests came back negative.

GENITALS: Which is a positive thing.

UNA: Blob Two is Genitals – Mum says they peaked in the sixties and refuse to retire.

GENITALS: It feels really nice to meet you in the flesh, Una.

UNA: You too, Ma'am.

HACK: Genitals are genderfluid. The go by 'they'.

UNA: I don't recognise this superhero. I can't see a protruding power, only a protruding knack of making me feel quite small.

HEAD: You won't have seen Hack on HumTV. She's our secret weapon, the Superhero of Change. She reverses those 'wrong' decisions we make for the civilians.

GENITALS: The rubber to our pencil. Or she's the spit and dab to the crumbs on our face.

HEAD: Hack is your port of call in error. So, Una, is it a yes or a no?

UNA: Um. I dunno. My dad's waiting outside so –

GENITALS: Thing is we don't need you to decide, we just need you to say yes.

UNA pulls up her top and bares her stomach.

UNA: Yes. Please take it. Give it to a superhero that knows how to handle it. I'll take anything else.

They look at me as if I just pulled an entire family out of my anus.

The door swings open and a silver-haired man flops in.

ROBBIE: Hum's motorway is gridlocked.

GENITALS: You're just in time Robbie, you got a knife? She wants us to cut out her gut!

UNA: The weedy man fiddles with his chest.

ROBBIE fiddles with his heart and becomes HEART.

HEART: Una, we are not organ thieves. We are the decision-makers that keep Hum humming.

HEAD: Demand is at its all time high. Decisions are needed all over Hum and we can't keep up. Hum needs something quicker. Something stronger. Something guttural.

UNA: They hover over me, taking up every inch of the room. Come on mouth, say something normal.

My mum's got you on our fridge.

HEART: Una, as a new recruit, you'll be a cog in the machine of helping civilians do the right thing for them.

UNA: Hack stares at me, her eyes change colour with every blink.

Me? Are you serious?

HEART: Will you join us, Una?

UNA: These superheroes are making the biggest decision of my life for me. To make *me* one of *them*? I can't save lives. I couldn't even decide what to say.

The lines on Head's forehead were bulging, Heart's chest pulsating, Genitals' flaps and folds jiggling. My eyes cross in disbelief.

GENITALS: Is that a yes?

HEAD: It'll be a long road. With responsibility, comes a platform. And with somebody as gifted as you, I suspect Hum will –

HEART: Fall in love with you. Don't tell anyone, not even your loved ones. I told an ex-lover once and she put it on Hum Forum – I had to move.

UNA: Head, Heart, I can't do this – I'm not – I can't – it's not me – I'm –

GENITALS: If you need to talk, there's always Maude.

HEAD: With you by our side, Hum will thrive.

HEART: I wish I had someone like you to call before I was made a superhero.

UNA: Come on Una. I plant my feet firmly and brace myself for the gate opening. The crowning. The big moment. And then I can go home.

HEART: We've been around the houses thinking of a name for you!

UNA: Heart hands me a medallion with 'H' on it.

HEART: 'Hunch'. We think its really nice to say out loud.

UNA: There's an awkward silence. Like parents waiting for their child to say her first word –
Yes.
They give me a thumbs up and usher me out of the door.
It's pitch black outside their office.

MAUDE: Well are you going to let me go home or do I have to stay and lick your feet?

UNA: I don't know.

7. THE FIRST SLEEPOVER

The night before the welcome ball. In MUM and DAD's place.

UNA: Luke.
Luke.
Are you awake?

LUKE: Hmmmmm

UNA: I had the dream again. The velcro nipple dream. Lefty's soft velcro and righty's scratchy velcro and they stick together, I go for a ten mile run, then when I'm finished, I rip the velcro apart but the velcro's vanished and they're just nipples of flesh again and I bleed out.

You're coming across quite side swappy. It's just that you decided on the right side after we – and then you keep sliding into the middle which throws sides out the window.

LUKE: Let's swap sides.

They swap sides.

UNA: This side's very warm.
Maybe you should go home. For tonight. And probably remain there. And not come here again.
It's just. I can't sleep. And I could say it's not you it's me, but it definitely is you because I have been known to sleep in the past otherwise I'd be dead.

LUKE: Is this about the – you know – the intercourse?

UNA: No! Yes. It happened, and I'm not sold on it. I wouldn't put my pocket money on us doing it again.

LUKE: It felt a bit –

UNA: Yes. No point discussing it. I tried it, kudos to me, but it's not going to be a new addition to the routine.

LUKE: Right. You do realise we didn't actually do it, though.

UNA: Yes. But the components leading up to it suggest it's not my bag. We set the table and I don't think I want the lunch.

LUKE: You know there's not really an order to it, or a conductor. It's just sort of two people treading water together until you both float.

UNA: Well I can only swim if there's no one else in the lane.

Knock at the door.

MUM: Is everything alright in there? Oh my Hum.

UNA: MUUUUMM.

LUKE: Thanks but I think I'm being kicked out –

MUM: Una. Is he real? Are you doing everything right?

UNA: MUUUUMM.

MUM: She's nervous about her first day tomorrow. Sounds exactly the same as the last job – but – you know our number monkey. If it's a fraction different, our Una…

LUKE: Well I actually am trying to get to know her a bit.

MUM: Aww – Una. *(Mouths a bunch of things that she thinks LUKE can't see.)*

UNA: Leave Mum!

MUM leaves.

Sorry.

LUKE: You know, when I get nervous about work stuff, I think about how small a significance I'm making. Not in a sad way. Yeah, I chef, I feed people – but if I disappeared, they'd just eat elsewhere. Or someone else would do it. The world would keep moving.

UNA: Yeah.

LUKE: Why don't we meet after your first day? Where?

UNA: I dunno.

LUKE: I'll sort it.

8. THE WELCOME BALL

At the Superhero Headquarters.

UNA: Huuuunch. HuNcH. HUncHH.

Maybe I need a stance. Or a quirk. Or a flaw. Something
that says 'I belong here' and not 'I found these legs in a bin'.
I check out my outfit in the mirror. They told me I had to
go red. I might never wear another colour again.

I hear hundreds of voices coming from the other end of
the corridor. Maybe I should do a runner. Maybe I could
knock my head on the wall and pass out, then they'd have
no choice but to find somebody else.

Sort it out Una. I adopt my gang walk I do sometimes
when I walk home late at night. Never been mugged.

*She opens the door and there's a cacophony of conversations being
had by the greatest minds in Hum.*

Inside the hall, everything looks like it doesn't live here
normally. Temporary swings, fountains and shrubbery. The
original building was burnt down by radicalists in the fifties.
The room is packed full of the who's-who of Hum. I creep
around the edge.

I spot the guy who cuts the ribbons. And the lady that
invented our currency that you can't hold.

A man with a tray carrying twelve different coloured
liquids approaches. The choice is sickening. I put my
blinders on. One step at a time, Una.

I pass a massive table of aging overweight men.

OLD MAN: All I'm saying is that Hum is going to the dogs, a
new inexperienced hire isn't going to change anything!
Gut decisions? About as accurate as throwing shit at a wall!
What are we calling her? Hornch? Hinch? She better shift
some merchandise.

UNA: I catch his eye, side smile and continue on. I regret not
asking my dad to wait outside.

I reach the other end of the room. My name plate is glistening. I'm next to Head. Heart. Opposite Genitals. Suddenly the music cuts – even the grey-haired money men shut up and stand up. Four people enter the room. There they are – Head, Heart, Hack and Genitals. Luminous. Mighty. And walking very slowly. The room erupts. I sort of half-stand, half-hover.

Superhero fanfare music.

I wonder if superheroes walk slowly so they don't sweat. I feel a hot breath on my neck.

GENITALS: I've made sixty-nine thousand decisions this week. Some fucking hot, some fucking middling, and some fucking bleak.

UNA: They sit down and neck an entire bottle of naughty liquid. Head and Heart embrace me at the same time. Polite. And passionate. Hack doesn't get up from her seat.

HEART: Don't mind her. She's going through a tough time. It's not personal.

UNA: Head takes the microphone.

HEAD: The people of Hum. Please take your seats.
As you well know, we've searched far and wide to find a new way of thinking corporeally, crossed all corners of Hum to discover a superhero that will rocket our metropolis into new heights.
We have found an exceptional individual, with great pleasure I introduce to you our new superhero, joining the ranks here with an extraordinary gift for gut intuition, it is an honour to welcome – HUNCH.

UNA: I pat my body for the scrumpled bit of – piss.

HACK: You can't go up like that.

UNA: Like what? Like what??

Genitals grabs my hands with their grubby mitts and makes me palm my stomach.

GENITALS: Go! Go!!

UNA: So often I have felt that I might keel over if someone blew on my face. But now, I feel weighted. Solid. Jenga stuck together with superglue.

Silence.

HUNCH: Whenever you feel in limbo, in flux, or in a rut.
I'll be at your beck and call with my gut.

UNA: I spoke in rhyme cos maybe that's what superheroes do.
But in reality I didn't have a fucking clue.
I see myself on a screen at the back of the room. It wasn't me. I'm huge, radiant, my gut stretching out like a golden lighthouse –

She inflates.

HUNCH: Hum is where I was born, where I grew up and where all my family are. I like it. And I'll make sure it likes me.

Pause.

HuuNNcHH!

The room erupts.

CROWD CHANT: HUNCH HUNCH HUNCH

HUNCH: I don't quite have a plan, the plan will come –

CROWD CHANT: HUNCH HUNCH HUNCH

HUNCH: I'll be a soft opening – so don't expect anything over night –

CROWD CHANT: HUNCH HUNCH HUNCH

HUNCH: I'll be there for you goddamnit, dusk 'til dawn.

UNA: Flash.

Flash.

Shallow breath.

Some hellos, some thank yous.

There's no 'getting tanked'. There's no 'what goes on in Superhero Headquarters stays in Superhero Headquarters' – a team of cleaners arrive at 10.15 and they've wheeled everything away by 10.30. Someone switches the last light off –

CLEANER: Sorry – can I – can I get a – (selfie).

HUNCH: Sure?

UNA: All the superheroes have their own twenty-four hour office. When you're last in, you get the one next to the loo. Superheroes shit alot.

Head shits tiny sporadic pellets and Heart monstrous feeling beasts. The quickest is Genitals – frequent but rapid like a woodpecker. *(Tongue trill.)*

MAUDE: Do you always listen to others defecate?

UNA: She took a deep breath as if the mere chore of talking to me was excruciating.

MAUDE: Have you got a next of kin? A partner, an ally, a someone that'll still love you when the world thinks your worthless?

UNA: Um… I don't know. Jools?

MAUDE: Joke. For some reason, in the eyes of a civilian, superheroes can do no wrong.

UNA: Right.

MAUDE: You should think about moving away from the centre of Hum.

UNA: Okay.

MAUDE: And all decisions, once you've made them, are logged in the Decidi-Bank. Can't access that unless you ask me. Oh. And one more thing –

UNA: – I've got quite a few questions actually –

MAUDE: I'm not a babysitter. If you need one of those, I suggest you resign now before you start to change the nation…

9. THE FIRST DAY

UNA: I had a holiday before I officially started as Hunch. They said to tie up loose ends and get fit. When I thought about the 'unknown' ahead of me the pain would return – so I ended up wasting time, I mean, spending time with my new boyfriend-thing-man-accompaniment-whatever-Luke.

I've become a bit fond of how he looks at me. He never puts a time limit on an answer, he just lets me be quiet, sometime an answer never comes and he just smiles. I like how he traces his finger left to right along my collarbone like he's removing dust from a ledge. Makes me feel delicate. And his hands are ginormous.

LUKE: I wasn't sure what project managers eat for lunch on their first day so brown and green. Una special.

UNA: I was too nervous to speak, let alone eat. I took comfort in wearing the trousers of my school uniform. Familiar. Practical. Itchy.

LUKE: I'm on the dinner shift tonight so I'll be back late.
You should start smoking, then you get breaks.
You'll be the best project manager/managing projector…
management ladyperson – ?
Are you leaving now or –

UNA: I don't know.

LUKE: That's okay – you'll get there. Knock 'em dead.

UNA: You too! Not the customers. Just the competition!

When he leaves it suddenly dawns on me I have *no* fucking idea how it happens. There's no pamphlet. Or gumph. Do I just roam the streets? Do I find a massive hill, stand with my legs wide open and spot the trouble from afar? Should I do home visits? Cold calling? Should I smell something in the air? Will I suddenly sprout breasts that know where I should be?

Can I fly? Can I hover? Can I crush metal with my eyes?

UNA jumps, trying to kick start her superpower. Nothing. Then, as if she's on top of a mountain.

Hunch here!
Ready-for-action?
Gut girl. Nanana.

MUM: Una? You want a mini yum?

UNA: I'm fine!

MUM: Fine, you don't want one, or fine you haven't decided yet?

She looks at her gut.

UNA: I feel an odd sensation in my stomach. Not the scorching pain I am used to, but a warmth, like a small sun is waking up inside me.

She puts a finger on her gut.

I wonder if I get a pension? Or dental care. Will I live forever?

She puts a full hand on her gut.

What if I get there and I destroy lives? Or it's a joke? And I'm on one of those HumTV shows that de-dignify stupid civilians.

Her gut pings. She swipes.

I'm in a room.

Not my room. A room I've never seen before. It smells like a handful of lemons has tried to give mould a hug.

I catch my breath sitting on a pile of clothes.

ALF: Don't touch the merchandise.

HUNCH: Oh! Sorry!

UNA: A very round man with what looks like a dozen jumpers on bats me off.

ALF: I'll turn the lights off.

HUNCH: No!

ALF: Your glow is hurting my eyes.

UNA: He waddles over to the dimmer.

HUNCH: You'll have to bear with me. This is my first time.

ALF: Does that mean you'll fuck it up?

HUNCH: No. They've put me on extremely low tier decisions for the first few months. So I'm not going to kill anyone.

UNA: He lets a bit of dribble run down his chin, and then his neck before he mops it up with one of the dozen jumpers.

HUNCH: Okay. So. What do you need deciding?

ALF: Should I take my jumper off or leave it on?

HUNCH: Is that it? Easy. *(Beat.)* Gut? The Power of Gut? Can you hear me?

23

ALF: Do you know what you're doing?

HUNCH hears his gut.

HUNCH: You should keep it on.

ALF: Thank you. Thank you.

UNA: He takes my hand and kisses it. I wonder how I'll ever wipe off the smell, but it also makes me feel brilliant.

ALF: You can stay if you like...?

HUNCH: No no no – just – working out how to leave.
Just. On. My. Wa –

HUNCH fiddles with her stomach for a beat too long until – swipe.

UNA: And that was it – I'm back in my parent's living room.
Until –

Swipe.

I'm looming over a cot. Less dizzy this time. More coherent.
Superheroes don't slouch.
Superheroes probably don't slur.
I adopt a new stance – subtle – but outstanding.

BABY: *(Baby noise.)*

HUNCH: Um. Put it in your mouth?

UNA: The baby grabs a brown bunny that definitely used to be white.

HUNCH: Oh wait – no – actually – floor – floor. I think about summoning Hack, but not sure how. Don't want to cause any trouble on my first day. I'll just leave it.

Swipe.

UNA: Back in my living room.

Swipe.

HUNCH: I'm somewhere familiar. I'm upright. I'm bold. I'm Hunch.

JOOLS: Hunch! Shit! You want one of these?

UNA: Jools!

Jools and I met at pre-school. Best friends ever since we went to hospital after drinking a potion we made from all the bits and bobs in her mum's bathroom.

HUNCH: Jools! Jools it's me! How mad is this?! Look!

JOOLS: Oh shit. Not normally this much of a tip.

HUNCH: Jools!

JOOLS: Do I tell Una that I kissed Dan's best friend or do I just sweep it under the rug?

HUNCH: Very good.

JOOLS: Which one?

HUNCH: Jools…

HUNCH waves.

JOOLS: Yeah?

JOOLS waves back. HUNCH realises she looks nothing like UNA.

HUNCH: Sweep it under the rug.

UNA: Sam comes into the room.

HUNCH: Sam!

SAM: *(Gasps in shock, then composes.)* I'm actually really busy rearranging my pegs.

UNA: She runs out the room. The back of her onesie has a tail that Jools has made from tampons.

JOOLS: World Book Day really crept on me this year. It was that or a cape made out of fag packets.

UNA: Jools had Sam when she was at Uni and although she's only six – no – seven now. She's got all of Jools' best bits without knowing all of the bad bits about me.

I wanted to stay but – first day and all.

She swipes.

HUNCH/UNA: Should I have that fifth pint of naughty liquid?
Should I copy my mate's homework?
Should I turn the tap off?

UNA: I made 192 decisions today. At midnight I try to recall what they all were but I only remember the thank yous. Feels like Hum is moving forward. That I've given a push to a couple of hundred people on a swing to start them off.

A pain. Ah. It's hunger.

I haven't eaten my lunch Luke made. Wonder if superheroes have special appetites.
That's the first time I've said superhero.
I'm a superhero.

10. THE HOUSEHUNT

UNA: Six months fly by, Luke met my parents properly, I met his parent, he got glandular fever for a bit, I got some moles removed during a lunch break, Jools likes him, his friends like me. It's understood that we love each other but neither of us have said it yet. So he – we – decide to move in. Together.

ESTATE AGENT: I can show you around or I can just leave you to it. Bedroom's in the living room, living room's in the kitchen. So – well, yeah. Any questions?

There's an animal noise from a cupboard.

That'll be taken care of. I'll leave you to have a minute whilst I get some non-fresh air.

UNA: We hear a 'fuck' as he trips up the stairs thanks to a broken bulb.

Sorry I was late. It's my boss. She's on my bloody back, riding me like a –

LUKE: I don't see much storage.

UNA: Luke wipes his finger along the surface of the stove. He didn't grow up with much and now, I'm fucking rolling in it, superheroes just get an endless payout. I can't tell Luke so I get out wads of cash from HumBank and leave them in the tip jar at his restaurant. He thinks it's for his oink belly burger.

Luke shared a room with his six siblings. I'm obsessed with all the caught red-handed masturbation stories. His youngest brother drunk all the naughty liquids a few New Year's ago and jumped off the Hum Bridge. So now, Luke isn't reckless, he just has a firm grasp of what is worth suffering for. And continuing to live with my parents wasn't one of them.

LUKE: I mean, it'll mean a different commute for you, but we can work on that. But it'll be ours.

UNA: To rent.

LUKE: Yeah.

UNA: I was going to save it, but, when Gran passed away she left me – well – a small fortune. We can use that?

LUKE: I don't know – that's yours –

UNA: Well, it was hers. It can be ours.

LUKE: I think it's better if it's actually ours.

27

UNA: Okay. Good idea.

He grabs me and kisses my forehead – again. I don't know why he keeps doing this. This is probably the optimum opportunity to say –

I love – but I spot a bunch of gunk coming out from the bathroom tiles – puss.

ESTATE AGENT: So – you in or out? Snatch it or be snubbed.

LUKE: Yep. We'll take it.

UNA: Really?

LUKE: Yeah.

UNA: I wasn't going to be in it much anyway.

So… I better get back to work – there's some wobbly data I need to 'put in the fridge' as they say.

LUKE: Right now? Really?

UNA: Yeah.

11. THE PHOTOSHOOT

HUNCH: So, sorry, what's this one for? Maude didn't say.

PHOTOGRAPHER: Um – not sure yet – could you just hold the python –

HUNCH: Yeah sure. Like this?

PHOTOGRAPHER: Bit more sure of yourself.

HUNCH: Course.

PHOTGRAPHER: Hunch you are magnificent! The camera wants to cook you breakfast in bed! Okay! Got it. Thanks Hunch. Just in case, what do you think of python?

HUNCH: I have no opinion.

PHOTOGRAPHER: For the press?

HUNCH: Oh, course, it deeply saddens me to think they are on their way out and um – delve deep into your pockets, if you can.

Snap. HUNCH looks directly into the camera.

12. THE MIDDLE TIER

UNA: Nowadays I can't wait to get to work. I high five Maude on the way in.

MAUDE: Don't touch my hand with your hand ever again.

Oh. And miraculously, you've passed probation so you've moved up a tier.

UNA: Amazing! Up a tier. That's not food and handjobs. That's careers and relationships. Life paths. Cool.
Maude farts as she walks away. Not an old person accidental fart, a deliberate, know who's actually your boss, fart.
I wait for my first summon like a bird of prey.

Ping. She swipes. Car park.

NELL: My boss chews like a small pig on a trampoline. Should I key his van?

HUNCH: Yes.

NELL is overcome with joy as she draws on the car.

Swipe. Outside a court room.

UNA: A man, nearly seven foot with arms as wide as I am long embraces me.

TOM: Thank you. Thank you so much for coming. Guilty or not guilty?

Are you not going to ask me which one is the lie?

HUNCH: Not guilty.

TOM: No, I did it. I want to apologise to her and her family and – I – if I say guilty at least it's honest, it's – the other lad's got off. James is on the bench, and Chris is back playing already.

HUNCH: Your gut wants you to say not guilty.

TOM: But I'm not! I'm not not guilty!

UNA: It's hard when their gut really clashes with the right thing.

Swipe. Houseboat.

I'm on a houseboat. A pixie-haired woman looks out the window. The boat is full of trendy streamlined toys. I can't hear anything. She's not asking me anything.

HUNCH starts to sign.

Crap – she signs. I don't sign.

She's having a conversation in sign.

I can sign!
Does she move home to Copenhagen or stay here?

UNA: Her gut wails at me. Home.

HUNCH: Home.

UNA: She gives me a massive hug and she bursts into tears.

SIGFRID goes. HUSBAND arrives.

HUSBAND: Where's Sigfrid? Where's my wife?

HUNCH: Fuck.

HUSBAND: You took her from me! You took her! You stole her! Jerry, Alex your mother's left us because of this –

HUNCH: He picks up an oar, slings it behind him and then – Sigfrid walks back through the door.

HUNCH signs: 'Why did you change your mind? Why are you back?'

SIGFRID signs: 'I don't know why I changed. I don't want to be here'

UNA: I left Sigfrid making a cup of tea for her husband that she clearly wants to leave. The oar still in his grip. Outside the window a pair of emerald eyes.

Swipe. Back with LUKE.

When I get home Luke's testing some new menu ideas.

LUKE: *(Frustrated.)* Arrghhh.

UNA: What?

LUKE: The yums. I keep thinking I want them round and then I change my mind and make them square. But I need them round!

UNA wafts her arms around the kitchen as if trying to catch a ghost.

What are you doing?

UNA: Nothing.

LUKE: Practicing our first dance?

UNA: For what?

LUKE: For Jools' wedding. Or…ours?

UNA: But we're not engaged.

LUKE: No. We're not.

UNA: What are the yums for?

LUKE: Work.

LUKE picks out a ring from the yums.

UNA: I can't deal with this right now and teleport to Hum department store.

13. THE HUNCH DOLL

UNA: I need new bras, I don't think they've got bigger because I'm a superhero but I don't want the teleporting to take its toll.

JOOLS: Stand there any longer Una and I'll try scan you at the till.

SAM: Dan's given Mum pocket money to buy something nice for the wedding.

UNA: I think she tries to wink but headbutts the air.
I know! We like Dan don't we?

SAM: Dan's fine for now. This one or this one?

UNA: She holds up two silky nighties.
They're both very…nice.
Feels weird complementing a seven-year-old on her choices of seduction.

SAM: Got to pick.

UNA: I don't know. What do you think?

SAM: This one. It's short and that means more arousing.

JOOLS: Sorry Una, don't know where she got that from – must be the telly. Now, how's things? You get my bridesmaid bollocks I sent on?

UNA: Yep. I'll reply when I'm back. I'm well – well, works… my boss…

JOOLS: Yeah Luke said.

> Ah – will you keep an eye on her for two ticks I want these puppies locked up under my chin.

JOOLS goes.

UNA: What you got there?

SAM: Hunch.

UNA: What?!

SAM: HHuuNNCccH.

SAM brandishes a small cuddly toy of HUNCH.

UNA: Wow! She's a lot more – rotund – than I imagined.

SAM: She's the best one because she's the reddest.

UNA: What would you ask her if you met her?

SAM: Do I have to call him 'dad'?

UNA: You don't have to ask Hunch that – you can ask me.

SAM: Do I have to call him 'dad'?

UNA: *(Can't decide.)*

SAM: Una! You don't know anything!

14. THE HUNCH FANS

In the flat.

UNA: Its a Hum national holiday and I'm taking it off.

> I've made round yums for me and Luke. And 'me' is Una. All day. And all night.

LUKE: Well, well, I won't say no…

UNA: Could wait for them to cool, or just have one now, or wait 'til tomorrow, or –

LUKE: Are you alright?

UNA smiles. A smile that's a bit too large.

Both rows of teeth and a bit of tongue – what the fuck have you done?! Wipe out a society?!
Your mum's worried you're being taken advantage of. Isn't project management all about delegation? That's what I do – I've got a bloke to pluck the clucks, a bloke to fry the moo, a bloke to…

UNA: It's not the same

LUKE: They share pretty similar –

UNA: They don't.

Beat.

LUKE: I spy with my little eye something beginning with 'U'.

UNA: Una.

LUKE: I wish. No.

UNA: Ulcer.

LUKE: You can't see inside my mouth.

UNA: I give up

LUKE: Unfair. What you're doing to me is really unfair.

UNA: He never looks at me when he confronts me. And has a habit of putting it into a seemingly fun but actually rubbish game.

LUKE: You hardly even brush your hair anymore and I'm not saying I need you to brush your hair – I just – love you and I can't love you if you're not around. Because then I can't see what I'm meant to be loving.

UNA: I took a day off. I'm here.

LUKE: I'm not a fucking eight-ball Una! Give me something, for fuck's sake?

UNA: I don't want this conversation. So I end it in quite an avante-garde way.

Sings a weird awkward exit song.

I bolt out to the garden we share with the other eleven flats. I pass the smokers. I always wish I was part of their club.

GROSS NEIGHBOUR 1: Going for a shag down the bottom, are we?

GROSS NEIGHBOUR 2: Yeah! Waiting for your fella al fresco frisko, eh?

UNA: Yep! Bonking by the bins! My favourite!

GROSS NEIGHBOUR 1: Did you hear about the explosion down South Hum? Warrants a ciggie – you want one?

UNA: Yes. No. I don't know.

I keep walking and stop at the pile of crap that everyone tries to get the bin men to take away.

FAN: She's here. She's here.

UNA: I peer over and there are two teenagers, leaning against the wall.

FAN: Does Hunch live here?

UNA: Who?

FAN: Hunch! Hunch! Hunch!

UNA: I looked to the left and Hum almighty, there are at least 200 teenagers skulking in the car park.
Um – I can see if she's about.

UNA ducks down, summons HUNCH.

HUNCH: Hello civilians.

FAN: OH MY HUM.

HUNCH: I have to keep saving Hum. But can everyone put their arms in the air? I want to feel like I'm flying!

HUNCH crowd surfs.

15. THE TIPPING POINT

HUNCH: It was clear Luke needed space so I threw myself into work.

Burn it.
Slap her.
Fire him.
Bluff!

HUNCH is asked for a selfie.

Crush them.
Beg her.
Bribe him.
Wank!

HUNCH is asked for an autograph.

Blame them.
Bag her.
Trick him.
Oh, these? Natural. I was born with them.

Mimic. Mask. Smother. Shun. Pray. Purge. Rave. Revolt!

UNA: As Una, I was getting abs.

HUNCH: As Hunch, I was unstoppable.

Swipe.

UNA: I'm on a single bed next to a figure in a hood.
He has a sprawl of video games littered around his feet
which means he could be anything from eleven to forty-one.

RIBS: I feel like I'm ridin' a boomerang.

HUNCH: It's a girl. What is your gut decision civilian?

RIBS: After the window smashed, everyone else did it,
everyone else took 'em but I didn't. I didn't take nothing, I
swear, I swear I didn't. But now I'm here with all this gear.

HUNCH: I'll report back to The Trunk, apologies for the hitch
in the system.

RIBS: It'll be on CCTV, my mum's gonna kill me, my mocks
start tomorrow.

HUNCH: Your gut says take them back.

UNA: She scuttles around the floor shoving the games into
a plastic bag. Something feels strange, like I am being
watched. The door creaks and I see a pair of walnut eyes
staring at me.

RIBS: Ah fuck, nah, nah, I think I should stay here.

HUNCH: Your gut wants you to return the games Rebecca.

HACK: Don't go! Stay.

UNA: My hands are white. I look around.

HUNCH: Hack! She's not asking you to reverse this, I've got this.

UNA: Her hair keeps changing colour, her body shape
morphing, a conveyor belt of shapes and textures.

HACK: Oh I know.

HUNCH: So why are you here?

HACK: We never get to chat. Must mean you're very good at your job. Are you happy?

HUNCH: Yes.

HACK: And Una. Is she happy?

HUNCH: Why does that matter?

16. THE HELP

In Superhero Headquarters.

UNA: I begged Heart to meet me for a mug of express liquid. He was off duty.

ROBBIE: *(Takes a sip.)* So. This is your office. What is it, fifteen sixteen square feet?

UNA: Yup.

ROBBIE: Yeah, mine has a south facing window. But uh, yours gets a lot of light.

UNA: Robbie, I'm not sleeping. Every minute I'm dolling out decisions quicker than I can even say them.

ROBBIE: And?

UNA: Is it possible Hack might be off form?

ROBBIE: Her partner was killed in accidental drive-by shooting six months ago. I don't think I'd be 'on form' – would you?

UNA: I tried the same thing that afternoon with Head.

HEAD: Sorry I'm late for this erratic and suspicious rendez-vous, Una. I'm here.

UNA: He was in superhero mode, excellent.

Head. My decisions are being reversed.

HEAD: Then you need to step up your game.

UNA: I think Hack might be –

HEAD: Hack has been here much longer than you, Una.

UNA: But the smallest of decisions can snowball into a complete ulterior life.

HEAD: That's the point. We change the world.

UNA: Right. We change the world. But what to?

HEAD: We take the baton of blame for Hum. We are the sole raconteurs of the world's narrative.

UNA: I can't continue at the pace I'm going. Hum is summoning me every second of the day – I can't –

HEAD: Hum has to keep humming. Civilians need us to decide –

UNA: Well then, I SHITTING QUIT.

HEAD: Look at me, Una. I know. You're not quitting. We're a team. We pick up where the others might drop. Keep up.

UNA: The Head downs his mug and leaves me.
There's a knock at the door.

She ducks down and hides.

GENITALS: Gut princess! You hear we are all on higher tier decisions now? Hum's on high alert limby! Life and death, honey! I'm going to get me a new hairdo!

17. THE HIGHER TIER

UNA: Hum was brimming with fires and floods. Life and death, honey.

In the vein of 2016, we hear snippets of news reports, a never ending list of disasters.

NEWS REPORTERS: Hum's youngest member of the police force has sadly passed away, found in his bedroom earlier this morn – an elderly couple have died after being stuck in – four school children have tragically passed – a bus, said to have had twenty-four passengers crashed into the side of a – a fire has broken out in South Hum – eighty-four fatalities, the latest count –

Swipe.

UNA: I stop in to the flat to get a few minutes shut eye.
Mum and Dad are in my bedroom.

DAD: It was good of her to organise something, unlike her to break her routine.

MUM: Unlike her to be late. I think she's having an affair Larry.

DAD: Una? She can't even have an affair with her school pants.

UNA: Before they can see me I head for the front door.
Luke's in the kitchen opening a bunch of cards with a bread knife.
It's his birthday. Bollocks.
I tiptoe out, praying he won't hear me, I'll make it up to him tomorrow.
I walk down the street and see Nathan.

NATHAN: *(On phone.)* Luke set up this shindig himself. Una's been rubbish apparently, is it crap to say I think he could do better?

UNA: I hide in a bush.

I can't do both. I piss off too many people.

Her stomach feels like it's caving in.

I think it'd be better for everyone if I say goodbye to Una.

18. THE WEDDING

UNA: The next day I'm bombarded by civilians but tear myself away to say goodbye properly. I leg it to the reception, I missed the ceremony, but I'm here. In the distance I see Luke leaning against the side of a barn.

LUKE: No, no she had to work.

BIX: On a Saturday?

LUKE: Yeah – she often works weekends. Do you want one?

BIX: You've none left!

LUKE: Oh yeah, can I have one of yours then?

BIX: You've such big hands!

LUKE: Oh, thanks, grew them myself. With my BODY!
So you're a bridesmaid or maid of honour – so many shades of blue.

BIX: Bridesmaid. Duck egg blue.

LUKE: Duck egg blue. Well it suits you – it sort of makes your eyes look like that colour too – when they probably already are – but you know.

BIX leans towards LUKE.

Listen – I'm not sure this is such a good idea – I do have a girlfriend –

GENITALS arrives.

UNA: *(Whispers.)* Genitals – what are you doing here?

GENITALS: So, you want to have a threesome with this guy and girl? Of course it's a yes.

UNA: No. Genitals. This is my boy-man-friend-thing. Has he summoned you?

GENITALS: No, I am here for the lady. Hey lady. Do it. You take it all off lady.

UNA: What if this was my out? To say goodbye for good. I know Luke, I can do no wrong in his eyes. It'll only end if he –

She swipes. She pushes GENITALS away.

HUNCH: His gut was saying no.

LUKE: You're nice, but I do have a girlfriend, she just isn't here – I –

Hunch? I didn't summon you

BIX: Oh my god Hunch! You're my favourite! Can I touch?!

HUNCH: You should sleep with – what's your name?

BIX: Bix.

HUNCH: You should take 'Bix' to your room and have guilt-free sex with her.

UNA: I watch Luke feel for his erection.

LUKE: I didn't ask for you.

HUNCH: What are you waiting for? Go on.

UNA: I watch them both stumble away. He looks around one more time, desperately hoping Una will save him from himself. And he's gone.

Genitals behind me, their breasts flinging into the sky.

GENITALS: I love weddings!

UNA: Inside the barn I see silhouettes of people twirling each other around. The occasional 'wooo' followed by the occasional 'woah woah – it's slip – aaah! hahaha!'
Fuck all those bending twirling bodies. Reeling on zero responsibility. 'Hunch made me do it!' Just one fuck off finger pointing right at me.

SAM: Hunch??

UNA: She's in a little blue glittery dress.

HUNCH: Off duty.

SAM: Can I ask you something??

HUNCH: Off. Duty. No.

UNA: I try to walk past her but she tugs on my suit.

SAM: I don't know what to ask!

HUNCH: Go back inside.

SAM: Um – should I go and get my Hunch doll?

HUNCH: Can everybody just make their own fucking decisions? Don't get it. DON'T GET IT.

UNA: She runs off and I transform back into Una.
Sam?? SAM?! Argh!
Need to set Luke free first.
'Luke. I want your stuff gone by the morning.'
I need a drink.
In the barn, I head to the liquids table and down the naughtiest liquid there.

JOOLS: Saved you a big plate of oink rolls in the kitchen.
Not sure I'm into this, Una. There are loads of people here I actually really dislike. I tell you, pot belly Paul from HR needs a bigger sized shirt. Ah! I love this one!

UNA: It was bittersweet watching Jools trying to do the splits as Una's last memory.

And Sam, under the table, kissing my ankles.

19. THE ANNOUNCEMENT

UNA: A few hours later, it's 6am. The superheroes have been called together but it's only Genitals in superhero form. Myself, Robbie and Jim are human, warts and all, scrappy hair and under eye bags.

MAUDE: Okay. I presume they'll just want some bitesize quotes they can use. Focus on the bright future. Have they seen our new fountain, for example.

UNA: Una didn't smoke.

UNA lights a cigarette.

My last cigarette as Una.

ROBBIE: This is truly a waste of time. Why are we here when we could be working?

JIM: Watch out for the bees.

ROBBIE: Say something of sense, for Hum's sake.

JIM: Bees don't have knees.

ROBBIE: I forgot how much I cannot stand you Jim.

UNA: I watch them both cross over to superheroes as footsteps sprint towards us.

LUKE: Excuse me. Excuse me – this is headquarters right? I want to make a complaint. An appeal.

MAUDE: Come back in our opening hours, civilian.

LUKE: No. I didn't summon Hunch. And she made me – she made me do something I – I want a Hack reversal.

UNA: Luke? In his wedding suit. It looks as if he's run through Hum backwards.

I try to change to Hunch, but he sees me.

LUKE: I knew it. I knew it was you.

UNA: What are you on about? I'm here to complain too.

LUKE: I knew it, I knew it – I would have never ever have done that. It has to have been you. You know me inside out, you know what that'd do – the afters – you –

UNA: You must still be drunk, Luke.

LUKE's phone is ringing.

LUKE: It's Jools. She keeps calling. I'm gonna get it.

Beat.

UNA: Four little chairs waited for us at a long table. Each with their own microphone. The crowd boo.

I'm Hunch only. Felt safe to be no one but Hunch.

MAUDE: *(Whispers.)* Okay – well you summoned this – you start it.

HUNCH: No I didn't. Head? Heart? Thought it was one of you –

HEAD: My wife was in the bath and I got a message saying I was expected.

HEART: I left my cat in the garden. I thought it was you.

GENITALS: My grandpappi is in hydrotherapy, I left him wrinkly to come here.

UNA: The crowd buzz, louder, angrier. Civilians cram into every space. Sardined, standing, arms in the air –

MAUDE: Fuck it. Quiet! I'll be taking questions for The Trunk?

AUDIENCE 1: Excuse me, excuse me, my oven keeps turning itself back on.

AUDIENCE 2: Me! Me! My hair is always wet – even though I'm always drying it!

AUDIENCE 3: I can't turn my tap off! It always turns back on! My house is ruin –

UNA: A figure parts the crowds and wades straight into the middle. A few gasps pepper the silence.

HACK: Ssssshhh. We need the room to be quiet. The superheroes are due a phone call and we won't be able to hear it, if we're all chatting.

GENITALS: *(Phone.)* No I told you to take him out of the pool, I remember, I did!

HEART: *(Phone.)* No no, she didn't drown, I saw the gardener turn the hose off.

HEAD: *(Phone.)* Judy? Who's this? No, no, no I told Judy to turn it off. I told her!

UNA: All three of them dart out the building whilst Hack slinks up to the microphone.

HACK: You see, whilst you've been battling with the bigger picture, slobbering over higher tier decisions, I've been babysitting the lower tier.
Every tap in Hum that was turned off, has been turned back on. Uh oh. It's an epidemic and it's been 'dripping' for a while now.

MAUDE: Everybody out. Hum please return to your homes.

HACK: Keep hydrated civilians.

MAUDE: Go! Turn off your taps.

HACK: But I have one more question, Maude. Hunch. Are you into watersports?

HUNCH: What have you done?

HACK: Scuba diving? You can decide later. It's just, Little Sam
needs picking up from the bottom of the pond.

HUNCH: What?

HACK: She was looking for the little version of you last night.
Not much of a swimmer, apparently.

UNA: Thirty-two missed calls. It's Mum. *(Phone.)* What? No
she didn't – I – what?
I didn't make her do it.
I can't remember what she asked me – did she ask me?
Maude? Maude – what did Sam ask me?

MAUDE: That's enough questions for today, thank you Hum.

UNA: And Hack was gone. Bursting with panic, the doors
jammed with civilians.
Sam.
Sammy.

MAUDE: I hate to ask. But are you going to cry because I'll
look the other way.

UNA goes to break but then – she swipes.

HUNCH: I'm fine. I need to see all the decisions made in the
last year Maude.

MAUDE: Cupboard next to the boiler. You need my pinkie.

UNA: She had colour-coordinated them. Head's, Heart's,
Genitals', a smaller file Hunch's and Hack's. Others made
by civilians themselves, right ones, wrongs ones, ones that
never mattered.
Hack. Formerly known as Sally, Superhero of Change.
Reversal of decision to tell partner Lola to wait in the
lobby. Denied.

Reversal of decision to tell partner Lola to meet her at work. Denied.

Hundreds and hundreds of decisions denied. All for Sally.

MAUDE: I liked Lola. Those shots weren't even meant for her. If she'd just waited somewhere else.

UNA: And why couldn't Hack reverse it?

MAUDE: She's been trying. I think that's why my sandals are soggy.

20. THE BATTLE

UNA: I run outside. I have to find Hack. I keep seeing flashes of Sam, wading into the pond. Her little blue dress getting darker and heavier.

I fly through summons, no (Hunch?), no (Hunch!), no, no, until –

Swipe.

HACK! She's standing on a garage door like it's a surfboard, brandishing a hose, watching an old man try to tread water.

HUNCH: Hack.
KcaH.
Hack!
KcaH!

HUNCH does breaststroke to get to HACK, but it's reversed immediately. She tries front crawl, it's reversed immediately.

Hack, leave the civilians. This is you versus me.

HACK: Never. Hum needs to stop humming and start dealing with its consequences.

UNA: She dives into the water, leaving no splash.

Swipe.

No, no, I can't find her, no, no, I check everywhere, then – out on a motorway that is now a river. I see Alf, the dozen jumper man trying to float.

ALF: Hunch! Hunch! Help me, should I take some jumpers off? Will I float better?

HUNCH: YES.

HACK: No.

HUNCH: YES.

HACK: No. Keep them on, doughball.

HUNCH: If you want to kill the whole of Hum, go ahead, but I'm going to have to kill you first.

HUNCH is the most 'superhero' she has ever been. HUNCH and HACK are no longer on stage, they are in a blockbuster.

I follow her onto a roof.

HACK: Didn't realise you were such a copycat, Hunch. I thought you made decisions for a living.

Combat swing.

HUNCH: I thought you reversed my mistakes.

Three combats.

Oops.

HACK: You're over Hunch. Go back to Una. Or has she undecided her way off a cliff?

Combat swing. HUNCH slips and hangs off the roof.

I've been cleaning up Hum's mess for years but nobody could help me clean up mine. Everyone must feel the isolating misery of losing a loved one because of their own mistake. Because of their human weakness.

HUNCH: That's not what your gut is saying.

HACK: Your gut talk doesn't work on me, Hunch.

HUNCH's gut starts to shine like a golden lighthouse.

HUNCH: At least if I die, I die true.

HACK: You'll die a glorified fortune cookie.

HUNCH: I'll die fighting for what's right.

HACK: Is that your final hunch?

HUNCH slips even further.

HUNCH: Your gut wants you to hide under a duvet.

HACK: What?

HUNCH: You want to be Sally.

HACK: I'd pat you on the back but I want you to live a few seconds longer.

HUNCH: It wants you to lay flowers on Lola's grave.

HACK lifts HUNCH up.

HACK: No one's said that name to me in months.

UNA: Her hair drops off in chunks, her skin darkens, her eyes stay chestnut brown. What's left is a woman at the first stage of grief. Sally.

SALLY: I'll reverse it, I'll reverse them all, if you'll come with me.

HUNCH: The five of us go back through every wrong reversal Hack had made.

HEAD: Sorry Judy, bit of a hitch, it's frightfully nice to have you back. Can I have a kiss?

HEART: You beautiful feline. Never leave me again.

GENITALS: Granpapi, think of all the grandmami's you'd leave behind.

HUNCH: And every civilian in Hum, even Alf…all restored. Hum was humming again.

SALLY: That's it. That's all of them.

HUNCH: You need to reverse Sam's.

SALLY: But I can't.

HUNCH: Yes, you can.

SALLY: That wasn't Hunch. If I reverse Hunch's decision, she drowns. It was Una's indecision. I'm sorry.

Flashback to the wedding.

UNA: I felt a tug from under the liquids table. Sam kissing my ankles.

SAM: Una? Should I go and get my Hunch doll or not? I asked Hunch. She said no.

UNA: Then don't.

SAM: What do you think?

UNA: I don't know! I really don't know!! Do you want it?

SAM: Yes.

UNA: Well…

UNA can't decide.

21. HUNCHELLA PART TWO

We return where we left off at the start of the play.

HUNCH: Fasten your seatbelts. It's Hunch.

Moments like this. This is what it's all about.

Thank *you* for letting *me* save *you*.

22. THE DRESSING ROOM

HUNCH enters the dressing room. She sits. She waits.

Silence.

There's a knock at the door.

STAGE MANAGER: Hiya – um – sorry, locking up in five, so if you could make your way out –

HUNCH: Yeah. Sure.

STAGE MANAGER: Eeeaaaaah, I so shouldn't be doing this, you're probably off duty but I'd shoot myself if I didn't – what should I make my boyfriend for dinner? Moo burger or cluck burger?

HUNCH: Did you fancy a glass of this with me?

STAGE MANAGER: Aaaah – he's waiting for me outside.

HUNCH: Yeah. Cluck burger.

STAGE MANAGER: Thank you! Locking up in four.

STAGE MANAGER leaves. HUNCH is alone.

There's a knock at the door.

HUNCH: Yup?

JOOLS: Sorry. I saw you on the telly in a shop.

What do I do?

What do I do?

What do I do without her?

What do I do?! Tell me what to do. Tell me. Do something.

JOOLS hits the floor. HUNCH can only mouth 'I don't know'.

23. THE MAKING IT RIGHT

UNA is in the car.

UNA: I start to spend most of my time in my car, as Una,
     practicing decisions.
     Left. Right. Stop. Go.
     I park outside Jools'.

MUM knocks.

MUM: Una? Una! What are you doing in there?!

UNA: Can I come back home with you guys after this?

MUM: Oh love, we're just not sure we can take the burden,
     pinkle.

UNA: The burden?

DAD: You're not a burden!

MUM: Well, we spoke to Luke. And – we – we love you very
     much Una, but we're unable to accommodate Hunch. If
     the neighbours found out. We can't condone a – well – a
     'superhero' –

DAD: We can send your brother round with money if you need
     money.

MUM: You should pay your respects quickly and then leave, love.

UNA: Standing in the corner of Jools' packed living room. The pitter-patter chatter was mostly weather based. Until –

FUNERAL GUEST 1: Apparently Hunch was spotted nearby the pond – did you hear? And she lives near here –

FUNERAL GUEST 2: Rancid piece of work. Could be someone here –

UNA: I was summoned, but I didn't accept it.
I was never going back to Hunch.
I clasped an overwhelming pain in my tummy.

JOOLS: Where the fuck have you been?
You expecting me to deal with this all by myself?

UNA: Her eyes were wide and her mouth was moving slower than her tongue.

JOOLS: Come on then. I'll introduce you to all my aunts that have come out of the woodwork because my kid's croaked before me. 'The most tragic of deaths'. Oh, this one has a moustache.

UNA: Jools. I –
But Luke wraps his arm around Jools and leads her away from me.

JOOLS: Naaaah – gotta bring Una. UNA! Come see the moustache! It's like a badgers come out of her nose!

UNA: I crawl out of the house and lie on the grass.

UNA's stomach is in immense pain.

NATHAN: Sis, did Hunch ever, you know, see me, you know, do stuff?

UNA: What stuff? He lies on the grass with me.

NATHAN: Nothing creepy, but like, did she ever see me on my last holiday to –

UNA: No. Different department.

NATHAN: Phew. Pretty cool though.

UNA: He kisses my cheek and scuttles off.

> Months pass and I never touch Hunch again after the
> awards. The entire Trunk disbanded. I get flowers from
> Head and Heart, no note. I saw Genitals in the Hum
> department store, they didn't look up.
>
> Living with Jools I'd practice decisions whilst she was in
> bed, try and make her eat something and then drive her to
> the grief counsellor's.
>
> One Tuesday she made me stop off.

JOOLS: Just go in, say how you feel, then come back out. I'll be
here. And Una – don't be all weird about it.

UNA steps into the restaurant.

UNA: I spy with my little eye something beginning with 'H'.

LUKE: Hunch?

UNA: No. No. No. Handsome. I think you are really handsome.

> Alright, my turn, you lost. I spy with my little eye
> something beginning with 'S'.
>
> Uh-Uh. You're not going to guess.
>
> Strapping. I think you're a strapping man. A big-handed,
> strapping man that can carry things, build things, fix things…
> Alright, your go.

LUKE: 'B'?

UNA: Um… Beautiful? No? Cool. B-B-B-B-B. BANANAS?
B-B-B-B-marriage material?

LUKE: Bitch.

UNA: Right. Fair. The marriage thing was a joke, you know.
We'd have to go out again first.

Heeelllooo? Shit. I thought spontaneous gestures after you fuck up were a guaranteed success – I thought that's why everybody does them in films. You can't even look me in the eye. Turns out even though you've been thinking about someone, every single day, every minute, it's clear they're not thinking about you anymore. And they don't even hate you, because hate would mean they still like you a bit, they just 'meh' you.

LUKE: Una, this in an open plan kitchen.

UNA: Gordon. Alf. Lorraine.
Okay. I'll let you get back to your cheffycheffy. Just one more thing though.
I spy with my little eye, something beginning with 'P'. Please?

LUKE: Please what?

UNA: Please accept my decision that I'm made up of decisions and indecisions. And I'll be wrong a lot of the time. And need advice sometimes. But they're all mine. I promise.

LUKE: It was you who kept filling the tips jar wasn't it?

UNA: No. Yes. I don't know.

THE END

WWW.OBERONBOOKS.COM